April 17, 2014

Dear Colleagues:

Please find enclosed a report entitled *Building a Brighter Future with Optics and Photonics,* prepared by the Fast Track Action Committee on Optics and Photonics (FTAC-OP) for the Physical Sciences Subcommittee of the Committee on Science, National Science and Technology Council.

Advanced optics and photonics are enabling technologies for applications in communications, computing, manufacturing, healthcare, energy, defense, and numerous other areas. Progress in the field of optics and photonics has the potential to generate new knowledge, promote economic growth, create new industries and new high-skilled jobs, and provide technologies for new applications.

The FTAC-OP developed a prioritized list of seven recommendations, which are organized into categories of research opportunities and research-related capabilities. The research opportunities address: (1) biophotonics to advance understanding of systems biology and disease progression; (2) faint to single photonics; (3) imaging through complex media; and (4) ultra-low-power optoelectronics. The research-related capabilities, which will encourage innovation in optics and photonics research, are: (1) accessible fabrication facilities; (2) exotic photonics; and (3) domestic sources for critical photonic materials.

Taken as a whole, the recommendations presented in this report are supportive of the recommendations of the 2013 National Research Council Report *Optics and Photonics: Essential Technologies for our Nation* as well as national priorities in biology, manufacturing, computing, and materials. The FTAC-OP strongly believes that support of these recommendations will accelerate basic research progress and applications in optics and photonics.

Sincerely,

The Committee on Science
National Science and Technology Council

About the National Science and Technology Council
The National Science and Technology Council (NSTC) is the principal means by which the Executive Branch coordinates science and technology policy across the diverse entities that make up the Federal research and development enterprise. A primary objective of the NSTC is establishing clear national goals for Federal science and technology investments. The NSTC prepares research and development strategies that are coordinated across Federal agencies to form investment packages aimed at accomplishing multiple national goals. The work of the NSTC is organized under five committees: Environment, Natural Resources and Sustainability; Homeland and National Security; Science, Technology, Engineering, and Math (STEM) Education; Science; and Technology. Each of these committees oversees subcommittees and working groups focused on different aspects of science and technology. More information is available at http://www.whitehouse.gov/ostp/nstc.

About the Committee on Science
The purpose of the Committee on Science (CoS) is to advise and assist the NSTC, under Executive Order 12881, to increase the overall effectiveness and productivity of federally supported efforts that develop new knowledge in the sciences, mathematics, and engineering (not including those areas primarily related to the environment and natural resources). The CoS addresses significant national and international policy, program, and budget matters that cut across agency boundaries and provides a formal mechanism for interagency science-policy development, coordination, and information exchange.

About the Physical Sciences Subcommittee
The Physical Sciences Subcommittee (PSSC) contributes to the activities of the Committee on Science (CoS) of the National Science and Technology Council (NSTC). The purpose of the PSSC is to advise and assist the CoS and the NSTC on U.S. policies, procedures, and plans in the physical sciences. As such, and to the extent permitted by law, the PSSC will define and coordinate Federal efforts in the physical sciences, identify emerging opportunities, stimulate international cooperation, and foster the development of the physical sciences. The PSSC will also explore ways in which the Federal government can increase the overall effectiveness and productivity of its investment in physical sciences research, especially with regard to issues that cut across agency boundaries. The PSSC comprises representatives from nine Federal agencies.

About the Fast Track Action Committee on the Optics and Photonics
The Fast Track Action Committee on "Optics and Photonics" (FTAC-OP) was formed in April 2013 under the auspices of the Physical Sciences Subcommittee (PSSC) of the National Science and Technology Council (NSTC). The FTAC-OP completed its work in September 2013. The FTAC comprised representatives from 14 Federal agencies. The purpose of the FTAC-OP was to identify cross-cutting areas of optics and photonics research that, with interagency, cooperation could benefit the Nation; to prioritize these research areas for possible Federal investment; and, as appropriate, to set long-term, outcome oriented goals for Federal optics and photonics research.

THIS DOCUMENT IS A PRODUCT OF
THE FAST-TRACK ACTION COMMITTEE ON OPTICS AND PHOTONICS
UNDER THE PHYSICAL SCIENCES SUBCOMMITTEE
OF THE COMMITTEE ON SCIENCE OF
THE NATIONAL SCIENCE AND TECHNOLOGY COUNCIL

Building a Brighter Future with Optics and Photonics

Table of Contents

Executive Summary 1

Recommendations

    (A1) Biophotonics to Advance Understanding of Systems Biology and
            Disease Progression 6

    (A2) From Faint to Single Photonics 9

    (A3) Imaging through Complex Media 11

    (A4) Ultra-Low-Power Nano-Optoelectronics 14

    (B1) Accessible Fabrication Facilities for Researchers 16

    (B2) Exotic Photonics 18

    (B3) Domestic Sources for Critical Photonic Materials 20

Appendix I: Charter for FTAC-OP 22

Appendix II: FTAC-OP Membership 25

FAST-TRACK ACTION COMMITTEE ON OPTICS AND PHOTONICS:

Building a Brighter Future with Optics and Photonics

Executive Summary

Optics and photonics are enabling technologies for applications in communications, computing, manufacturing, healthcare, energy, defense, agriculture, and numerous other areas. For example, with improved detection and analysis of photons, deeper views into biological systems could be realized, earth and battlefield imaging enhanced, and long distance communications extended. New optoelectronic devices that range in size from a few microns to tens of microns could increase the speed of networks and computers dramatically, while significantly reducing power requirements. Optogenetics, where genetically altered neurons become sensitive to light, causing them to become active or suppressed upon illumination, offers a powerful tool for brain research.

Progress in the field of optics and photonics has the potential to promote economic growth, increase agricultural productivity, create new industries and new high-skilled jobs, and provide technologies for new applications, such as capable exascale computing and neurophotonics. Thus, there has been growing interest by academia, industry, and government in examining U.S. optics and photonics research and development to identify and prioritize research opportunities.

Towards this end, the Physical Sciences Subcommittee (PSSC) chartered the Fast-Track Action Committee on Optics and Photonics (FTAC-OP) in the spring of 2013 to respond to the National Research Council's (NRC) report, *Optics and Photonics: Essential Technologies for our Nation*, which was published in 2013[1]. The charter for the FTAC-OP is reproduced in Appendix I. Specifically, the PSSC requested that the FTAC-OP identify opportunities for Federal investment and interagency cooperation in basic and early applied research in optics and photonics.

The FTAC-OP met weekly over the course of its 120-day charter, inviting subject matter experts from government, academia, and industry as well as representatives of leading scientific societies to brief the Committee. Experts were asked to look forward to the next decade of technology development, and identify areas of basic and early applied research that will be important for advancing optics and photonics research and applications. Based on these briefings, the FTAC-OP has developed a prioritized list of recommendations, which has been organized into two categories, research opportunities and capability opportunities. The first category directly addresses the goals of the FTAC-OP, while the second category addresses gaps in research-related capabilities that are limiting discovery and innovation in optics and photonics research. The FTAC-OP's recommendations are presented in Box 1.

The set of recommendations categorized above address either directly or indirectly four of the five overarching grand-challenge questions put forth in the NRC report (see Box 2). The FTAC-OP did not attempt to directly address the NRC's solar-energy grand challenge question (Question 4 in Box 2) since the Department of Energy (DOE) is already investing significantly in this area through the SunShot Initiative and other programs.[2] It is also important to note that, while the NRC questions were developed to highlight *technological* gaps in the advancement of national priorities and competitiveness, the FTAC-OP recommendations focus on identifying

---

[1] National Research Council (NRC), *Optics and Photonics: Essential Technologies for Our Nation* (Washington DC, National Academies Press, 2013). For further details see http://www.nap.edu/catalog.php?record_id=13491.
[2] The Department of Energy's SunShot Initiative is described at http://www1.eere.energy.gov/solar/sunshot/index.html.

> **BOX 1**
> **Prioritized Recommendations of the**
> **Fast Track Action Committee on Optics and Photonics**
>
> *Research Opportunities*
>
> (A1) **Biophotonics to Advance Understanding of Systems Biology and Disease Progression.** Support fundamental research in innovative biophotonics to enable advances in quantitative imaging; systems biology, medicine, and neuroscience; *in vivo* validation of biomarkers that advance medical diagnostics, prevention, and treatment; and more efficient agricultural production.
>
> (A2) **From Faint to Single Photonics.** Develop optics and photonics technologies that operate at the faintest light levels.
>
> (A3) **Imaging Through Complex Media.** Advance the science of light propagation and imaging through scattering, dispersive, and turbulent media.
>
> (A4) **Ultra-Low-Power Nano-Optoelectronics.** Explore the limits of low energy, attojoule-level photonic devices for application to information processing and communications.
>
> *Capability Opportunities*
>
> (B1) **Accessible Fabrication Facilities for Researchers.** Determine the need of academic researchers and small business innovators for access to affordable domestic fabrication capabilities to advance the research, development, manufacture, and assembly of complex integrated photonic-electronic devices.
>
> (B2) **Exotic Photonics.** Promote research and development to make compact, user-friendly light sources, detectors, and associated optics at exotic wavelengths accessible to academia, national laboratories, and industry.
>
> (B3) **Domestic Sources for Critical Photonic Materials.** Develop and make available optical and photonic materials critical to our Nation's research programs, such as infrared materials, nonlinear materials, low-dimensional materials, and engineered materials.

basic and early applied research and capability opportunities in optics and photonics that will provide the foundations for bridging these technological gaps. The committee strongly believes that support of its recommendations will accelerate basic research progress and applications in optics and photonics.

The FTAC-OP recommendations are broadly enabling, contributing to important technology opportunities, national priorities, and key areas of the economy. For example, the highest priority research need, Biophotonics (Recommendation A1), and Imaging through Complex

> **BOX 2**
> **National Research Council (NRC) Grand Challenge Questions from**
> *Optics and Photonics: Essential Technologies for our Nation*[1]
>
> (1) How can the U.S. optics and photonics community invent technologies for the next factor-of-100 cost-effective capacity increases in optical networks?
>
> (2) How can the U.S. optics and photonics community develop a seamless integration of photonics and electronics components as a mainstream platform for low-cost fabrication and packaging of systems on a chip for communications, sensing, medical, energy, and defense applications?
>
> (3) How can the U.S. military develop the required optical technologies to support platforms capable of wide-area surveillance, object identification and improved image resolution, high-bandwidth free-space communication, laser strike, and defense against missiles?
>
> (4) How can U.S. energy stakeholders achieve cost parity across the nation's electric grid for solar power versus new fossil-fuel-powered electric plants by the year 2020?
>
> (5) How can the U.S. optics and photonics community develop optical sources and imaging tools to support an order of magnitude or more of increased resolution in manufacturing?

Media (Recommendation A3) are foundational to the President's BRAIN initiative[3] and the National Bioeconomy Blueprint[4] as they will enable new optics and photonic tools for imaging, controlling the electrical activity of the brain, and mapping protein and protein interactions at the cellular and subcellular level to advance systems and synthetic biology.

The highest priority capability opportunity, also the committee's highest priority overall recommendation[5], Accessible Fabrication Facilities (Recommendation B1), together with the Nano-Optoelectronics (Recommendation A4) research opportunity, provides the optics and photonics research and research capabilities to respond to the NRC's leading grand challenge question on achieving the next factor-of-100 cost-effective capacity increases in optical

---

[3] BRAIN: Brain Research through Advancing Innovative Neutrotechnologies. For further details see http://www.whitehouse.gov/infographics/brain-initiative.

[4] The National Bioeconomy Blueprint is described at http://www.whitehouse.gov/sites/default/files/microsites/ostp/national_bioeconomy_blueprint_april_2012.pdf.

[5] The recommendation responds to a recurring theme that the committee heard from the numerous scientists who briefed the committee that the need for inexpensive access to fabrication capabilities to enable state-of-the-art basic and early applied research on integrated photonic-electronic devices. The committee did not have the time or resources required to make specific recommendations addressing this need and instead recommends that a detailed study be undertaken to better understand what is required by the research community and how these requirements might best be met.

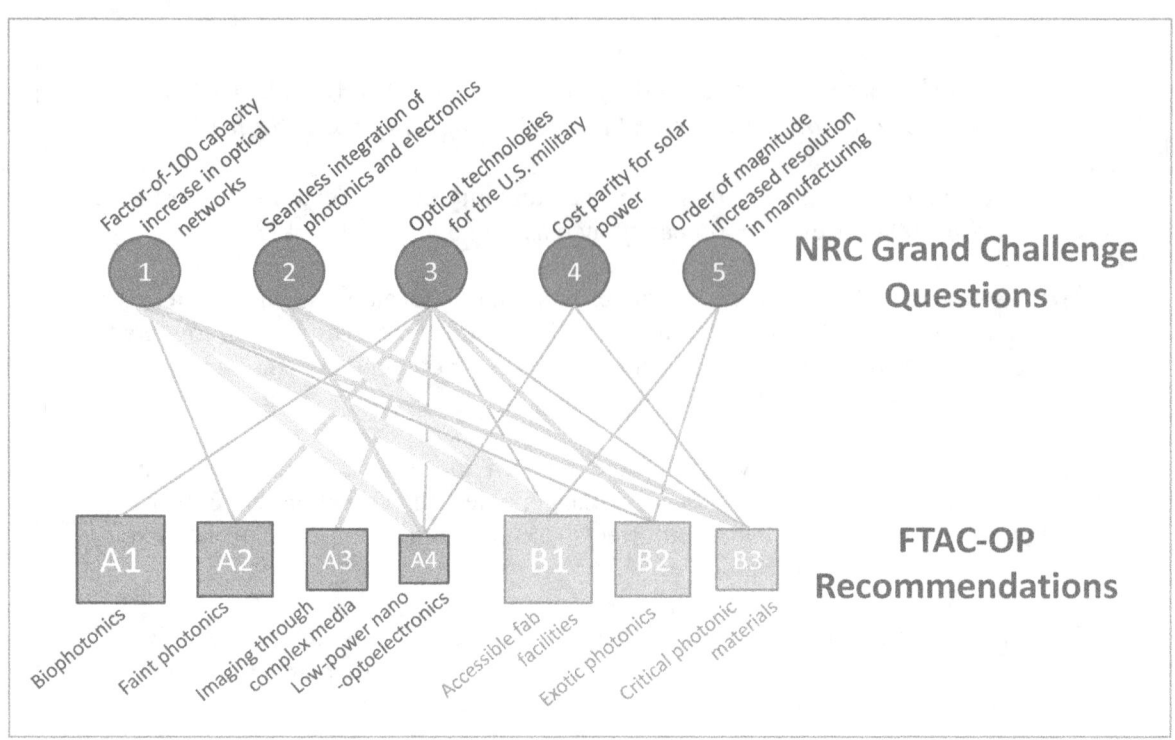

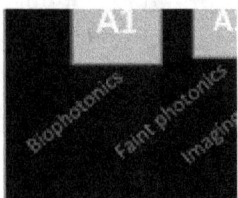

Mapping of the FTAC-OP recommendations the NRC's Grand Challenge Questions (top) and national priorities (bottom). The relative sizes of the boxes indicate the FTAC-OP prioritization, and the relative thickness of the connecting lines indicates the strength of the relationship.

networks. This capacity increase would contribute over the next decade to continued U.S. technology innovation and allow the realization of growing opportunities in Big Data. These two recommendations are also key enablers for achieving high-end capable computing at the exascale[6,7] by working towards eliminating bottlenecks in communications in multiprocessor, parallel computing environments and by reducing the energy costs and waste heat associated with moving information. Additionally, the Accessible Fabrication Facilities recommendation enhances opportunities for government-industry-academia collaboration in an area of national priority, advanced manufacturing.

The above research and capability recommendations depend upon the success of the FTAC-OP's recommendations on the development of compact, affordable, turnkey sources and detectors available at exotic wavelengths (Recommendation B2) and the development and availability of optical and photonic materials (Recommendation B3). The ability to easily generate exotic wavelengths for laboratory experiments will provide researchers across many fields with new tools and methodologies for the investigation of the structure of nanoscale materials, while accessible and validated sources of materials will help advance basic and applied research dependent on these materials.

Together, if implemented, the FTAC-OP recommendations will contribute to advancing innovation, important to U.S. economic growth and competitiveness. Federal support of these recommendations will foster a healthy, viable optics and photonics research and innovation environment with an impact felt by laboratory researchers, small business innovators, manufacturers, and consumers[8]. Furthermore, these recommendations will support Administration priorities to advance government-industry-university partnerships; promote innovation and commercialization of Federal research and development investments; develop a highly-skilled, high-wage earning labor force; provide an adequate pipeline of science, technology, engineering, and mathematics (STEM) workers at all levels; and create new industry sectors.

The following pages provide detailed descriptions of the FTAC-OP's seven recommendations.

---

[6] The Networking and Information Technology Research and Development Program: Supplement to the President's Budget, FY2014. See http://www.whitehouse.gov/sites/default/files/microsites/ostp/nitrd_fy14_budgetsup.pdf

[7] "Exascale Computing Technology Challenges," by J. Shalf, S. Dosanjh, and J. Morrison of DOE available at http://www.nersc.gov/assets/NERSC-Staff-Publications/2010/ShalfVecpar2010.pdf.

[8] Although, the recommendations do not directly address areas of opportunity such as quantum information, solar energy, optical computing, and chemical detection, the advances enabled by the FTAC-OP recommendations will benefit a broad range of basic and applied research areas dependent on optics and photonics.

(A1) Biophotonics to Advance Understanding of Systems Biology and Disease Progression

Recommendation: Support fundamental research in innovative biophotonics to enable advances in quantitative imaging; systems biology, medicine, and neuroscience; *in vivo* validation of biomarkers that advance medical diagnostics, prevention, and treatment; and more efficient agricultural production.

Motivation: Currently, there are many areas of biomedical research in which optics and photonics make important contributions. Monitoring tissue health and detecting disease (e.g., cancer detection and diagnosis, atherosclerotic plaque identification, glucose sensing, etc.) require the development of new diagnostic technologies and an understanding of the information they provide. Quantitative imaging, for example, offers the opportunity to measure early response to therapy. In drug trials or clinical practice, this would provide information on drug performance and patient response at an early time point, permitting physicians to offer more precise therapy by adjusting the regimen of therapy to be specifically tailored for individual patients.

Biophotonics is a critical field for the realization of the goals of the BRAIN initiative[2]. For example, progress in optogenetics, where genetically altered neurons become sensitive to light, causing them to become active or suppressed upon illumination, will be an indispensable tool in brain research and mapping the functional connectome of the brain. Improvements are needed in biomarker targets such as opsins as well as in DNA transfection methods and optical instrumentation that would greatly expand the potential for clinical optogenetics.

Often, medical devices do not detect the disease itself, but the effects of the disease on tissue (e.g. cancer can cause an increase in certain blood serum proteins, a change in tissue fluorescence at the cancer site, or a greater uptake of glucose in the tumor). These effects, called biomarkers, allow optical methods to be used as indirect indicators of the presence of disease. Many health issues can be better understood through improved understanding of the immune system and how it responds to threats of disease. The use of endogenous and exogenous biomarkers along with *in vivo* or point-of-care optical detection methods will enhance disease prevention and early detection. In addition, optical methods in conjunction with appropriate biomarkers offer the potential to reduce the time required for pharmaceuticals development, safety testing, and efficacy testing.

Progress in biomedical research is measured by the advancement of ideas and methods from laboratory studies to clinical utility. Thus, it is imperative that regulatory science issues important to the Food and Drug Administration (FDA) be addressed well before clinical translation is attempted. Research on optical biomarker validation must include efforts to characterize biocompatibility and photo damage (e.g., mechanisms, thresholds) and to develop rigorous, standardized test methods for safety and performance testing, if successful commercialization is to be achieved.

Agriculture faces the challenge of maintaining national and global food security while meeting escalating demands for food, fiber and fuel with increasing limited quality water resources and heightened competition for water use. Progress in optics and photonics could offer enhanced crop management (precision pesticide, fertilization, irrigation applications) at the farm scale level. This would enable producers to improve crop water productivity, limit water and fertilizer wastage from run-off and deep percolation, and reduce agricultural nonpoint source pollution.

Implementation Strategy and Agency Roles: The Federal Government has the opportunity to promote research on
- *in vivo* subcellular and sub-organelle imaging of proteins, membrane potentials, and signaling/messenger molecules and how changes serve as biomarkers for disease and/or drug interactions;
- the optical properties of materials on the molecular scale and nanoscale to make brighter, targeted probes for structural, molecular, and functional imaging;
- biophotonics areas linked with the BRAIN Initiative[2], such as full-field, deep, spatially-specific excitation of neurons and full-field, deep monitoring of neuronal activity;
- the relationship between phenotype, as detected by imaging and photonic interrogation of tissue, and the underlying genotype;
- regulatory science issues such as photodamage, biocompatibility, and standardized testing of performance and safety; and
- high-throughput phenotyping for selection of plant hybrids that are resistant to drought, disease, and pestilence in step with adaptation to global climate changes.

The highly focused nature of Federal laboratories and funding agencies may unintentionally limit the exchange of information between the physical/engineering sciences and the life sciences across the Federal Government. Federal agencies with extramural funding programs can facilitate a process of exchange, for example, by:
- establishing regular meetings between program managers representing the physical/engineering/mathematical disciplines and the life sciences agencies,
- promoting information sharing through temporary program manager exchanges, and
- increasing joint extramural funding opportunities between agencies.

In addition to facilitating exchange between Federal agencies, efforts should be initiated to develop high-quality interdisciplinary research teams organized to solve translational research problems encompassing laboratory to clinic issues. For example, efforts should
- encourage interdisciplinary teams that bring physical, optical, mathematical, and biological scientists and engineers and informatics specialists together for collaborative interactions, and ensure that applications from these newly formed teams have an avenue for unbiased peer review (e.g. the establishment of review panels with sufficiently broad expertise across all the relevant disciplines to render a review on each application for research brought to them),
- create funding opportunity announcements for academic and industrial participation that support the design, development, and validation of biophotonic methods to test safety and efficacy of new pharmaceuticals for disease detection and therapy,
- establish laboratory and clinical research projects and programs that feature optical technologies in pursuit of fundamental biological questions, and
- develop training opportunities specifically targeted at enhancing multidisciplinary expertise.

Federal agencies with intramural laboratories should also facilitate information exchange. For example, this could be accomplished by:
- providing sabbatical opportunities for exchange of scientists between Federal laboratories,
- creating a web broadcast seminar series across agencies to acquaint agency personnel to the issues of biophotonics,
- developing joint intramural programs and facilities, and
- encouraging participation of other agencies in laboratory peer review.

Success Measures:

- Expanded understanding of biological processes and the commercialization of optical technologies that will reveal these processes.
- Increased disease detection, especially at the early stages of development.
- Enhanced education and training of the workforce in biophotonics research.

Impact: Federal support in the area of biophotonics will help enable breakthroughs in optics technologies, including disease biomarker discovery and the understanding of brain functions, cell signaling, and genetics, potentially leading to a reduction in the impact of major diseases such as diabetes and cancer. These breakthroughs will:
- stimulate translational efforts involving both academic research institutions and industrial institutions, including small business activities, as ideas and processes are moved through the translational pipeline toward FDA approval and clinical utility, and
- help to fulfill the missions of the National Institutes of Health (NIH) and National Science Foundation (NSF), while benefiting other government agencies.

(A2) From Faint to Single Photonics

Recommendation: Develop optics and photonics technologies that operate at the faintest light levels.

Motivation: Basic research areas in the controlled generation, manipulation, and measurement of a few to single photons will directly promote advances in multiple fields including high-speed, secure, and energy-efficient communications; optical and photonic approaches to classical and quantum information processing; and detection and imaging at the faintest light levels. The smallest detectable unit of light is a single photon. Manipulation and detection of light at the single photon level will enable the maximum extraction of information from optical systems. For example, in remote sensing one could detect from longer distances or with higher resolution; in long distance space communications one could communicate over longer-distances and use significantly less power; and in biological imaging one could see deeper into tissue and extract more information from the sample as each photon is detected. In many applications, it is not practical or feasible to use a brighter source of light, and these applications would benefit from better optical components and techniques that work efficiently at few or single photon levels.

Implementation Strategy and Agency Roles: Based on the committee's findings, the government has the opportunity to
- promote coordination of research efforts among agencies in basic research areas to advance the development of photonics that works at a few to single photon levels,
- encourage formal and informal meetings among Federal funding agencies to discuss research direction,
- encourage coordination of investments in Small Business Innovation Research (SBIR) and Small Business Technology Transfer (STTR) programs to promote the transition of developments in basic and applied research in faint photonics to commercial products,
- promote the development of low-loss components including both bulk optics and photonic light wave circuits,
- pursue the development of higher quantum efficiency (approaching 100%) detectors and the capability to achieve this enhanced performance at non-cryogenic operating temperatures,
- advance the development of single photon sources on demand and other non-classical sources of faint light, and
- support the development of related infrastructure to enable development of faint photonics such as compact cryogenics for detectors, turnkey laser systems, and measurement technologies for characterizing components.

Success Measures:
- Development of devices and techniques that quantitatively measure and controllably produce light with quantum resolution and high accuracy.
- Use of single photonics to address needs in communications, sensing, imaging, and metrology.

Impact: Every photon contains information. If we could reliably manipulate light at its faintest levels, we would be able to communicate farther and with less energy, image with greater resolution, and see farther into the distance or into diffusive materials. Detection at the lowest levels of light would also mean we could reduce the intensity of light needed to probe biological samples, allowing higher spatial resolution measurements without potentially damaging the samples. Furthermore, measurement of light at the lowest levels possible has wide-ranging applications from imaging light from other galaxies to detecting neural activity in the human brain to detecting trace chemicals remotely.

(A3) Imaging through Complex Media

Recommendation: Advance the science of light propagation and imaging through scattering, dispersive, and turbulent media.

Motivation: Light scattering, dispersion, and aberrations in media such as biological tissue, the atmosphere and clouds, ocean water, and smoke critically limit the performance of optical techniques. These phenomena degrade the resolution, intensity, and spectral quality of detected signals, which reduce effectiveness in fields as diverse as oncology, agriculture, remote sensing and meteorology, homeland security and defense, and manufacturing.

The impact of biophotonic technologies has been most significant in applications involving superficial or transparent tissues (e.g., retinal imaging). However, for major health care issues involving deeper tissue sites (e.g., breast cancer), rapid degradation in signal quality with depth limits the utility of diagnostic and therapeutic approaches. This effect also represents an impediment to optical imaging in fundamental biology and pharmaceutical research involving small animals. While progress is being made with coherence-based, multi-modal, and nonlinear microscopy techniques, the ability to provide high resolution imaging centimeters deep into soft tissue and bone would have widespread impact. Furthermore, optical devices that provide selective optogenetic control or phototherapeutic effects at large depths would represent powerful new tools for the laboratory and clinic.

Advances in techniques for modeling and reducing optical aberrations (e.g., adaptive optics) will impact astronomical imaging through improved quality and reduced costs. In Earth resource mapping, the ability to image vertically through clouds increases operational efficiency and mission effectiveness. Additionally, military and civilian applications are greatly enhanced by the ability to maintain free-space communication; rapidly detect biological and chemical agents; conduct surveillance and reconnaissance operations through fog, smoke, vegetation, and clouds; and image horizontally in turbulent air. Regional-scale evapotranspiration, land-use estimates, and crop management for sustainable agriculture will be greatly enhanced by the ability to dependably calculate vegetative indices.

Miniaturized and low-cost computational imaging systems would have application as vehicle, equipment, or aerial mounted optical instruments, with algorithms that serve to detect the percent of vegetative cover and the presence and percent of disease, its severity, and its spread. Such information is necessary to provide decision support for precision crop management on a temporal scale.

Providing these capabilities will require further elucidation of scattering processes through the development of theoretical methods and computational tools. Advances in radiative transport theory, statistical inverse theory, numerical inversion methods, simulation models, and hybrid imaging models would play key roles. There is also a need for advanced optical instruments to serve as a foundation for the development of novel, high-penetration-depth systems, including (a) robust, high power, turnkey sources with broad bandwidth and/or wide tunability; (b) stable, low-noise, high-sensitivity detectors; (c) specialized devices such as rapid, high precision micro-mirror arrays for adaptive optics, and (d) computational imaging systems that integrate complex mathematical models and optical design to optimize detection of specific features or attributes. Miniaturization and cost reduction will accelerate adoption of these technologies into other fields.

Implementation Strategy and Agency Roles: Based on the Committee's findings, the Federal Government has the opportunity to promote research on light propagation in heterogeneous media in three key areas:
- theoretical and computational methods for analyzing, understanding, and optimizing light propagation and imaging in heterogeneous media,
- low cost, high performance optical instrumentation such as sources, detectors, and unique, device-specific components, and
- novel, high-quality imaging technologies with greater penetration capabilities in heterogeneous media (esp. those with relevance to specific agency missions).

To maximize impact and avoid duplication of effort on a topic where there is widespread government interest, a mechanism should be established to improve communication, collaboration, and sharing of optical imaging research among agencies and to promote collaboration and interaction between the public and private sectors. For example, as a resource for academia and industry, the government could work with professional societies to establish a web-based forum for U.S. researchers to exchange ideas, form collaborations, and share computational modeling codes and data.

The inherent appeal of imaging represents an opportunity for building a workforce for optics and photonics. For example, secondary and post-secondary educational and research training programs in optical imaging could be established. By incorporating sound mathematical and scientific preparation, these programs would have long-term positive impact the Nation's STEM workforce.

Success Measures:
- Groundbreaking theoretical and computational approaches to better understand, simulate, and optimize light propagation are realized.
- Low-cost, high-performance, optical components and systems that enhance imaging in/through heterogeneous media are developed.
- The number of publications and patents and the capabilities of optical imaging scientists entering the optics workforce are increased and the optical imaging industry offers new products leveraging advances in light propagation through heterogeneous media.

Impact: Progress in light propagation science will directly impact a wide variety of national interests, including national security and defense, astronomy, agriculture, the environment, and public health. It may help improve patient care through detection and treatment of diseases such as cancer, and will have an important role in the Federal BRAIN Initiative[2]. A range of Federal departments and agencies may benefit from new mission-critical imaging technologies and research tools, including the Department of Defense (DoD) (treaty monitoring; strategic and tactical military operations), Department of Homeland Security (surveillance for border security), National Aeronautics and Space Administration (astronomy and remote sensing), and

FDA (safety studies), as well as DOE, National Institute of Standards and Technology (NIST), Environmental Protection Agency, NIH, and U.S. Department of Agriculture.

The items described in this recommendation will help to retain world-class optical researchers and top students in the US despite ever-increasing international competition, leading to domestic advances in light propagation science with implications across numerous fields from cellular biology to astronomy. These innovations should drive the emergence of startup companies and benefit the wide variety of industries that rely on optical technology. The maintenance of a well-trained workforce in optics will increase the legacy of research investments and enhance the pool of scientific talent for academia and industry.

(A4) Ultra-Low Power Nano-Optoelectronics

Recommendation: Explore the limits of low-energy, attojoule-level, photonic devices for application to information processing and communications.

Motivation: Continued scaling of information processing systems to higher performance and lower power consumption may require that optical linkages or interconnects rather than electrical interconnects be used to move information at increasingly shorter distances, such as for on-chip and between-chip communications for multiprocessor chips. Conventional microelectronics-based computing is increasingly limited by power requirements and on-chip and between-chip communication bottlenecks. Such limits currently constrain information systems at all scales, from portable to exascale computing.

Although electronic logic requires only femtojoules for switching, it can require up to three orders of magnitude more energy for communication, even for length scales as small as a silicon chip. Additionally, electronic wiring severely limits bandwidth density even at the chip-scale. In contrast, the energy required for optical transmission will not inhibit the scaling of devices because it is largely independent of distance, and high information densities are possible in optical channels. Emerging optoelectronic devices that range in size from a few microns to tens of microns are now demonstrating communication energies on the order of 10 femtojoules per bit. Such optical energies are not close to fundamental limits for optics. By exploiting nanoscale structures, the potential exists for combined electronic-optoelectronic device structures that perform both logic and communications with attojoule energies and truly transform future information processing and communications.

Implementation Strategy and Agency Roles: The government has the opportunity to promote research on ultralow power nano-optoelectronics. Specific areas of opportunity are:
- fundamental limits of low energy devices,
- performance enhancement by sub-wavelength components, nanoscale semiconductor and metallic structures, and quantum confinement in quantum wells, wires, or dots,
- nanophotonic waveguides, resonators, and slow-light or non-periodic structures for light concentration and guiding,
- index of refraction engineering and new materials,
- nanoscale growth and fabrication approaches,
- intimate integration with active electronic functionality, such as transistors, and
- novel optoelectronic architectures and logic.

Success Measures: Ultra-Low Power Nano-Optoelectronics would lead to the following:
- Devices and systems with reduced size, weight, power consumption, and improved performance parameters over current state-of-the-art, benefiting signal processing, communications, and information systems.
- Advances in electromagnetic, electronic device simulation tools and circuit design tools.
- New types of light sources using metallic and semiconductor nanostructures.
- New detector designs in which metal performs a light concentration and charge extraction function.
- Advances in nanoantennas and nonlinear plasmonics.

- Lower energy requirements for signal processing and communication, which are core requirements for achieving the growth in the performance of information systems needed to meet future demands. (The cost of electricity has become a significant factor in information systems. Energy consumption is a new performance metric that will drive electronic-photonic integration.)

Impact: Support of this recommendation will help lay the foundation for the creation of ultra-low-energy optoelectronic devices. Such devices will enable smaller, lower weight, and lower power components; increase the capability, dependability, and survivability of future aerospace, satellite, and defense platforms; advance low-energy information processing and communication; and accelerate progress towards capable exascale computing. Furthermore, advances in ultra-low power nano-optoelectronics will likely be critical for achieving the next factor-of-100 cost-effective capacity increase in optical networks.

(B1) Accessible Fabrication Facilities for Researchers

Recommendation: Determine the need of academic researchers and small business innovators for access to affordable domestic fabrication capabilities to advance the research, development, manufacture, and assembly of complex integrated photonic-electronic devices.

Motivation: Increasingly sophisticated photonic-electronic integrated circuits and systems are expected to play a critical role in answering the first two of the five grand-challenge questions posed in the NRC report (see Box 2). U.S. innovation in silicon photonics is presently limited by the lack of access by academic researchers and small businesses to affordable, U.S.-based, commercial fabrication facilities capable of manufacturing complex integrated photonic-electronic devices. Thus, researchers are forced to use offshore fabrication facilities in Asia and Europe, where they risk loss of U.S. intellectual property while helping to grow non-U.S. intellectual property. University-based fabrication facilities do not have the design, modeling, and fabrication tools and reliable processes required to fabricate complex devices, and the cost to obtain and maintain such capabilities in a university environment is prohibitive. The high cost and long delay presently experienced by U.S. researchers in fabricating photonic-electronic devices and circuits slows the pace of innovation by favoring lower-risk incremental advances over higher-risk revolutionary advances in device designs.

The continued use of overseas fabrication facilities by U.S. researchers enables non-U.S. entities to become the leaders in the design, modeling, and fabrication of silicon photonic devices. This, in turn, encourages continued growth of overseas, high-skilled jobs in silicon photonics manufacturing and assembling and in the associated supply chains.

Open foundries, such as MOSIS[9], have been critical for allowing academic researchers and small "fabless" companies to develop a wide variety of applications in CMOS-based integrated electronics. Analogous open foundries in silicon and hybrid photonics could help promote U.S. research, development, and commercialization in integrated photonic-electronic technology by providing affordable and convenient access to experimental fabrication facilities to test new device designs, fabrication approaches, and materials.

Implementation Strategy and Agency Roles: Both government and the private sector have the capability to undertake the proposed assessment. The assessment should consider the following:
- the ability for researchers and small businesses to gain affordable access to domestic and foreign, commercial fabrication capabilities,
- the research and economic impact of researcher-accessible, domestic commercial foundries and custom fabrication facilities to
    - maintain U.S. competitiveness in the development and manufacture of integrated photonic-electronic devices,
    - pursue projects that address DoD needs, including the need for secure fabrication capabilities in silicon photonics,
    - broaden the community of researchers advancing the state of the art in integrated photonic devices,

---

[9] MOSIS: Provides Multi-Project Wafer (MPW) fabrication services in microelectronics. For further details see http://www.mosis.com/.

- establish a viable pipeline of skilled workers, and
- encourage formation of small, high-tech startup companies, broadening the innovation and application base for the technology,
- the research and economic impact of the loss of U.S. intellectual property to foreign entities,
- the need for associated design, modeling, assembly tools and standardized manufacturing processes to aid researchers and innovators,
- training opportunities for community college and university students and veterans, and
- strategies to implement any resulting recommendations.

The assessment should consider the experiences of national and international multi-project wafer foundries, such as MOSIS, OpSIS[10], and ePIXfab[11], and any recommendations from the full-day NSF *Workshop on US-Based Silicon Photonics Foundry/Fabrication Resources*[12].

Success Measures:
- Assessment completed and available to the public.
- Strategies proposed and considered for implementing any recommendations from the assessment.

Impact: The ability for researchers and small businesses to gain access to domestic, commercial fabrication capabilities could accelerate innovation and help enable the U.S. optics and photonics community to invent the technologies required for the next factor-of-100 cost-effective capacity increases in optical networks. Additionally, securing the domestic commercial foundry and custom fabrication facilities for researchers may help maintain U.S. competitiveness in the development and manufacture of integrated photonic-electronic devices, broaden the community of researchers advancing state of the art of integrated photonic devices, and establish a viable pipeline of skilled workers. Implementation of this recommendation will help the Administration determine the Federal Government's role in supporting domestic fabrication capabilities.

---

[10] OPSIS, or Optoelectronic Systems Integration in Silicon, is a multiwafer-project foundry capability described at http://opsisfoundry.org.
[11] ePIXfab is a foreign foundry service described at http://www.epixfab.eu.
[12] Workshop was held at the headquarters of the Optical Society of America in Washington, D.C on September 4, 2013.

(B2) Exotic Photonics

Recommendation: Promote research and development to make compact coherent sources, detectors, and associated optics at exotic wavelengths and to make them accessible to academia, national laboratories, and industry.

Motivation: Exotic wavelengths, such as those at the wings of the visible and near-infrared spectrum, are potentially powerful probes of chemical and physical processes in basic research and applications. Currently, generation of light at these wavelengths requires either specialized facilities, such as synchrotron radiation facilities, or high cost, custom-built hardware and instrumentation. These circumstances limit the accessibility of exotic wavelengths to researchers and stifle the pace of inquiry, innovation, and applications development. Extending the generation of coherent radiation to wavelengths outside of the visible and near-infrared and making the sources compact, affordable, and easy-to-use will open new areas of research and discovery. For example, terahertz and gigahertz sources and detectors have applications in imaging, spectroscopy, medicine, communication, and astronomy, whereas x-rays are powerful probes of the nanoworld. Exotic light, depending on wavelength, can penetrate thick objects and image small features in three dimensions. Spectra can be recorded with elemental and chemical specificity, and provide information on dynamics relevant to function. Compact, user-friendly x-ray sources, for example, may lead to disruptive technology in areas such as coherent diffractive imaging of single particles and nanolithography. Further basic research and metrology is needed to understand how to generate, manipulate, and control radiation at these exotic wavelengths, and Federal and non-Federal support should be assessed to determine whether it is sufficient to maintain U.S. competiveness with large centers in Europe and Asia[13].

Implementation Strategy and Agency Roles: The Federal Government has the opportunity to promote basic research at x-ray, extreme ultraviolet, mid- and long-wavelengths, and terahertz wavelengths and beyond to realize the following:
- turn-key, user-friendly, compact sources and associated components, such as detectors and optics,
- better tools and techniques for metrology at these wavelengths,
- increased experimental and theoretical research in ultrafast, x-ray science,
- new academic facilities in x-ray science to broaden the base of the user community, and
- an expanded training network for students and postdocs similar to ATTOFEL in Europe.[14]

---

[13] For example: Attoworld (http://www.attoworld.de/) and CFEL (http://www.cfel.de/) in Germany, CNR-IFN (http://www.ifn.cnr.it/home) in Italy, SACLA (http://www.lightsources.org/facility/sacla) in Japan, and LPC (http://laserspark.anu.edu.au/index.php) in Australia.
[14] ATTOFEL is a Marie Curie Initial Training Network funded by the European Union intended to advance the study of ultrafast dynamics using attosecond and XUV free electron laser sources (http://www.attofel.eu/).

Success Measures:
- Growth in the number of installed compact x-ray sources, and in their capabilities.
- Extent of new discoveries made using photonics at exotic wavelengths.
- Level of progress made toward scale-up of technologies for commercialization.
- Commercial benefits realized from investments in basic research at extreme wavelengths.

Impact: A wider availability of compact sources, components, and detectors at exotic wavelengths will lead to increased innovation and productivity in source development and in the basic and early applied research enabled by these sources.

(B3) Domestic Sources for Critical Photonic Materials

Recommendation: Develop and make available optical and photonic materials critical to our Nation's research programs, such as infrared materials, nonlinear materials, low-dimensional materials, novel fiber-optic materials, and engineered materials.

Motivation: Advances in engineered, sub-wavelength, and nanostructured materials are enabling innovative optics and photonics research and applications. Progress, however, is slowed by the difficulty that researchers have in gaining timely access to critical and innovative materials that are affordable and of high quality. Additionally, the DoD needs stable and reliable sources for defense-critical optical and photonic materials. Many of the key optical and photonic research materials are either manufactured overseas or produced by individual research groups. The growth, processing, production, and characterization of these materials often require significant expertise and highly specialized equipment not available to many research groups. Research groups with the capability to produce these materials may not be willing or able to share samples with other research groups due to competition, costs, or lack of production capacity. For overseas suppliers, the quality of key materials often varies from one acquisition to the next. Consequently, research groups either abandon efforts involving difficult-to-obtain materials, or begin making, purifying, and characterizing their own materials. Many research groups, however, often lack the measurement capabilities to assess the quality of homemade or purchased materials.

Implementation Strategy and Agency Roles: The Federal Government has the opportunity to promote the development of U.S. supplies for critical optical and photonic materials, research on new materials, and the advancement and dissemination of the metrology required to assess and quantify material performance and quality. These recommendations are aligned with the goals of the Materials Genome and with the two recommendations on "Strategic Materials for Optics" highlighted in the NRC report[1]:

- The U.S. R&D community should increase its leadership role in the development of nanostructured materials with designable and tailorable optical material properties, as well as process control for uniformity of production of these materials.
- The United States should develop a plan to ensure the availability of critical energy-related materials, including solar cells for energy generation and fluorescent materials to support future LED development.

To realize this opportunity, the FTAC-OP recommends developing a prioritized list of strategic optical and photonic materials required by U.S. basic and applied research programs. The inventory should highlight the quantities needed, supplier gaps, available quality levels relative to requirements, and whether research into alternative or promising new materials and methods to quantify material quality is needed. This inventory would help the government:

- promote the development of US-based suppliers,
- encourage suppliers to develop production methods that are documented, reproducible, and able to produce materials of high quality,

- encourage universities and small businesses to provide researchers with state-of-the-art materials not available commercially,
- facilitate access to present and develop new measurement capabilities, such as at NIST, to assess product quality and to provide feedback on quality to producers,
- explore alternatives for critical materials only available from non-US suppliers,
- promote research into new materials, materials properties measurement, and materials property databases, including on laser-materials interaction, which will support continued U.S. innovation in optics and photonics, and
- expand U.S. capabilities and university access to optical fiber fabrication capabilities for nanostructured and photonic band-gap fibers.

Success Measures:
- High quality and affordable optical and photonic materials that are critical to U.S. research and development are available within the U.S.
- Innovative materials with new and novel properties are developed and commercialized.

Impact: Research innovation and productivity will be advanced due to increased availability of high-quality, state-of-the-art optical and photonic materials. Commercial applications dependent on these materials will be accelerated. University suppliers will train students in materials manufacturing and characterization, provide a trained workforce for domestic materials suppliers, and facilitate technology transfer between academia and industry. University faculty and former students will be able to start new companies to manufacture critical optical and photonic materials. Federal support of this strategy will spur domestic production of key optical and photonic materials, provide validated sources of these materials for national defense and security applications, and accelerate research progress in optical and photonics materials and applications.

APPENDIX I

CHARTER
of the
FAST-TRACK ACTION COMMITTEE ON OPTICS AND PHOTONICS
PHYSICAL SCIENCES SUBCOMMITTEE
COMMITTEE ON SCIENCE
NATIONAL SCIENCE AND TECHNOLOGY COUNCIL

A. **Official Designation**

The Fast-Track Action Committee on Optics and Photonics (FTAC-OP) is hereby established by action of the National Science and Technology Council (NSTC), Committee on Science (CoS), Physical Sciences Subcommittee (PSSC).

B. **Purpose and Scope**

The 2012 National Research Council (NRC) report *Optics and Photonics: Essential Technologies for our Nation* offers a set of challenges and recommendations for realizing the potential of optical and photonic science. The purpose of the FTAC-OP is to (1) identify cross-cutting areas of optics and photonics research that, with interagency cooperation, could benefit the Nation based on the challenges and recommendations described in the 2012 NRC report; (2) prioritize these research areas for possible Federal investment; and (3) as appropriate, to set long-term, outcome oriented goals for Federal optics and photonics research.

These efforts should result in a prioritized list of optics and photonics research opportunities of national interest. The FTAC-OP may also recommend a mechanism or mechanisms for coordination of interagency collaboration and may produce other work products (e.g., joint requests for proposals, schedules) that support the committee's purpose.

C. **Functions**

The functions of the FTAC-OP are to:

1. Building on the findings of the 2012 NRC report, identify research areas of individual agency interest; opportunities for collaboration among agencies; and mechanisms for interagency coordination and collaboration.
2. Report the activities and progress of the FTAC-OP to the PSSC.
3. Make prioritized research and research organization recommendations to the PSSC.

D. Membership

The following NSTC departments and agencies are represented on the FTAC-OP:

- Department of Commerce (Co-chair);
- Department of Defense;
- Department of Health and Human Services;
- Department of Energy;
- National Aeronautics and Space Administration; and
- National Science Foundation (Co-chair).

The following elements of the Executive Office of the President shall also be represented:

- Office of Management and Budget; and
- Office of Science and Technology Policy.

Cooperating departments and agencies shall include other such Executive organizations, departments, and agencies as the co-Chairs may, from time to time, designate.

E. Private-Sector Interface

The FTAC-OP may seek advice from members of the President's Council of Advisors on Science and Technology to secure appropriate private sector advice and will recommend to the PSSC and/or the Director of Office of Science and Technology Policy the nature of any additional private-sector advice[1] needed to accomplish its mission. The FTAC-OP may also interact with and receive *ad hoc* advice from various private-sector groups as consistent with the Federal Advisory Committee Act.

F. Termination Date

This charter shall terminate 120 days from the date of inception, unless renewed by the PSSC Co-chairs.

G. Determination

I hereby determine that the establishment of the Fast-Track Action Committee on Optics and Photonics is in the public interest in connection with the performance of duties imposed on the Executive Branch by law, and that such duties can best be performed through the advice and counsel of such a group.

---

[1] The Federal Advisory Committee Act, 5 U.S.C. App., as amended, does not explicitly define "private sector", but the phrase is generally understood to include individuals or entities outside the Federal government such as, but not limited to, the following: non-Federal sources, academia, State, local or tribal governments, individual citizens, the public, non-governmental organizations, industry associations, and international bodies.

Approved:

signature 4/23/2013
Patricia Dehmer   Date
Co-chair of Subcommittee on Physical Sciences and
Deputy Director of the Office of Science
Department of Energy

signature 4/23/13
Gale Allen   Date
Co-chair of Subcommittee on Physical Sciences and
Acting Chief Scientist
National Aeronautics and Space Administration

signature 4/18/13
Fleming Crim   Date
Co-chair of Subcommittee on Physical Sciences and
Assistant Director for Mathematical and Physical Sciences
National Science Foundation

APPENDIX II

FAST-TRACK ACTION COMMITTEE ON OPTICS AND PHOTONICS
PHYSICAL SCIENCES SUBCOMMITTEE
COMMITTEE ON SCIENCE
NATIONAL SCIENCE AND TECHNOLOGY COUNCIL

Co-Chairs
Clark Cooper
Senior Advisor, Science in MPS
National Science Foundation

Gerald Fraser
Chief, Sensor Science Division
National Institute of Standards and Technology

Executive Secretary
Catherine Cooksey
Research Chemist, Optical Radiation Group
National Institute of Standards and Technology

OSTP Liaison
Gerald Blazey
Assistant Director, Physical Sciences
Office of Science and Technology Policy

Members
Ravindra Athale
Program Officer
Office of Naval Research

Amir Gandjbakhche
Senior Investigator, National Institute of Child Health and Human Development
National Institute of Health

Viktoria Greanya
Chief, Basic Research
Chemical and Biological Technologies Department
Defense Threat Reduction Agency

Dai Hyun Kim
Associate Director
Office of the Secretary of Defense

Jeffrey Krause
Program Manager
Department of Energy

Prem Kumar
Program Manager
Defense Advanced Research Projects Agency

William Luck
Deputy Chief Engineer
National Aeronautics and Space Administration

Joseph Mait
Senior Technical Researcher
U.S. Army Research Laboratory

Robert Nordstrom
Program Director, National Cancer Institute
National Institute of Health

Sae Woo Nam
NIST Fellow, Quantum Electronics and Photonics Division
National Institute of Standards and Technology

Joshua Pfefer
Laboratory Lead
Food and Drug Administration

Gernot Pomrenke
Program Manager
Air Force Office of Scientific Research

www.ingramcontent.com/pod-product-compliance
Lightning Source LLC
Chambersburg PA
CBHW081314180526
45170CB00007B/2705